MX

Fly

Barrie Watts

A & C Black · London

Here is a fly.

Have you ever seen a fly like this one?
It is a greenbottle.

In the summer, greenbottles buzz around
looking for sweet things to eat.

The greenbottle has found a left over slice of
bread from a picnic.

This book will tell you how a greenbottle comes
from a tiny egg.

A CIP catalogue record for this book
is available from the British Library.

ISBN 0–7136–3214–3

Published by A & C Black (Publishers) Limited
35 Bedford Row, London WC1R 4JH

Acknowledgements
The artwork is by Helen Senior.
The publishers would like to thank Michael Chinery for his help and advice.

Filmset by August Filmsetting, Haydock, St Helens
Printed in Belgium by Proost International Book Production

The greenbottle has big eyes.

The greenbottle has huge eyes. Each eye is made up of lots of little eyes. The greenbottle can see all around itself. It can see objects and other insects which move quickly.

Look at the small photograph. The greenbottle is walking upside down on the ceiling. The greenbottle grips on to the ceiling with sticky pads on its feet.

The greenbottles mate.

In the summer, the male and female greenbottles mate.

The male climbs on to the female. They stay together for up to one hour.

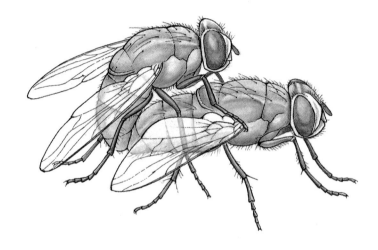

The female greenbottle flies away to look for a dead animal to lay her eggs on.

The female greenbottle has antennae on the front of her head. She uses the antennae to pick up the smell of a dead mouse.

The female greenbottle lays her eggs.

The female greenbottle lays about 150 eggs on the dead mouse.

Look at the photograph. The greenbottle's eggs look big. But in real life each egg is as small as the top of a pin.

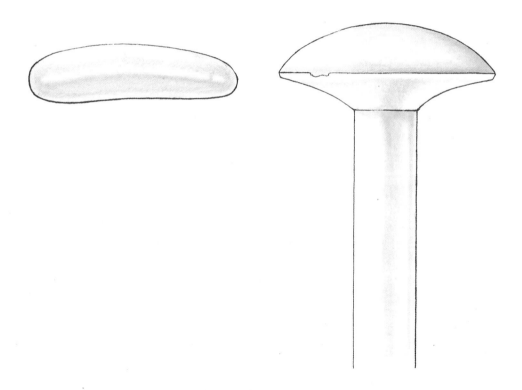

The eggs hatch.

The next day, tiny maggots come out of the eggs. They do not like sunlight. Look at the big photograph. The maggots bury into the mouse's fur and look for food.

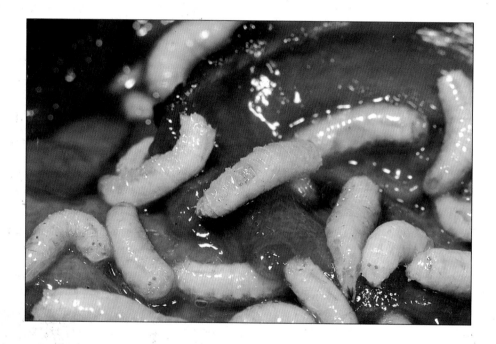

They turn the mouse's flesh into liquid.

The maggot's mouth is on the pointed end of its body. The maggot eats the mouse by sucking up the liquid. This liquid makes the maggot grow into a fly.

The maggot digs a hole.

The maggots eat the mouse in a week. They leave only the skeleton and fur.

After a week the maggot stops eating. It is fully grown. The maggot is as big as an apple pip.

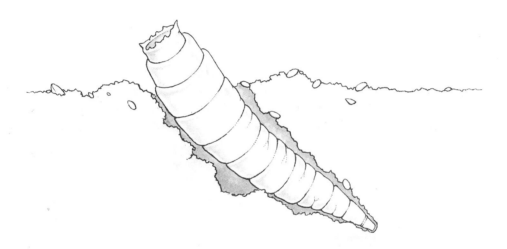

Look at the photograph. The maggot is made up of segments which help it to move. The maggot wriggles away from the mouse's skeleton and underneath the top layer of soil.

The maggot turns into a pupa.

In these photographs the maggot has been put on top of the soil so that you can see it change into a pupa.

Two days after burrowing into the ground, the maggot's skin starts to harden.

After another day, the maggot's skin is brown and hard. Look at the small photograph.
Inside the maggot's skin, a pupa is growing.

After three more days, the skin is black.
Inside, the pupa is growing into a greenbottle.

The greenbottle comes out of the pupa.

In a week, the greenbottle comes out of the pupa shell. Look at the big photograph. The greenbottle uses its head to break out of the pupa shell.

The greenbottle crawls away to a safe place on a leaf. Its wings are wet and crumpled.

The greenbottle waits for its wings to dry.

After ten minutes, the greenbottle's wings are full size.
But the greenbottle cannot fly yet because its
wings are still soft.

The greenbottle sits on the leaf and waits for its
wings to dry.

The greenbottle flies away.

After fifteen minutes, the greenbottle's wings are dry. It flies away to look for food.

In an hour, the greenbottle's body will be completely hard and green.

The greenbottle sleeps during the winter.

The greenbottle only flies when it is warm. In the
winter, most greenbottles die, but some find safe
places to stay. Look at the photograph.
The greenbottle has found a hollow place in a tree.

In the spring, the greenbottle will fly away.
It will look for a mate. What do you think
will happen then?

Do you remember how a greenbottle came from an egg?
See if you can tell the story in your own words.
You can use these pictures to help you.

1

2

4

5

Index

This index will help you to find some of the important words in this book.

In the summer, watch a greenbottle flying around. Look for the greenbottle's compound eyes.